This Math workbook belongs to

Numbers

1 One

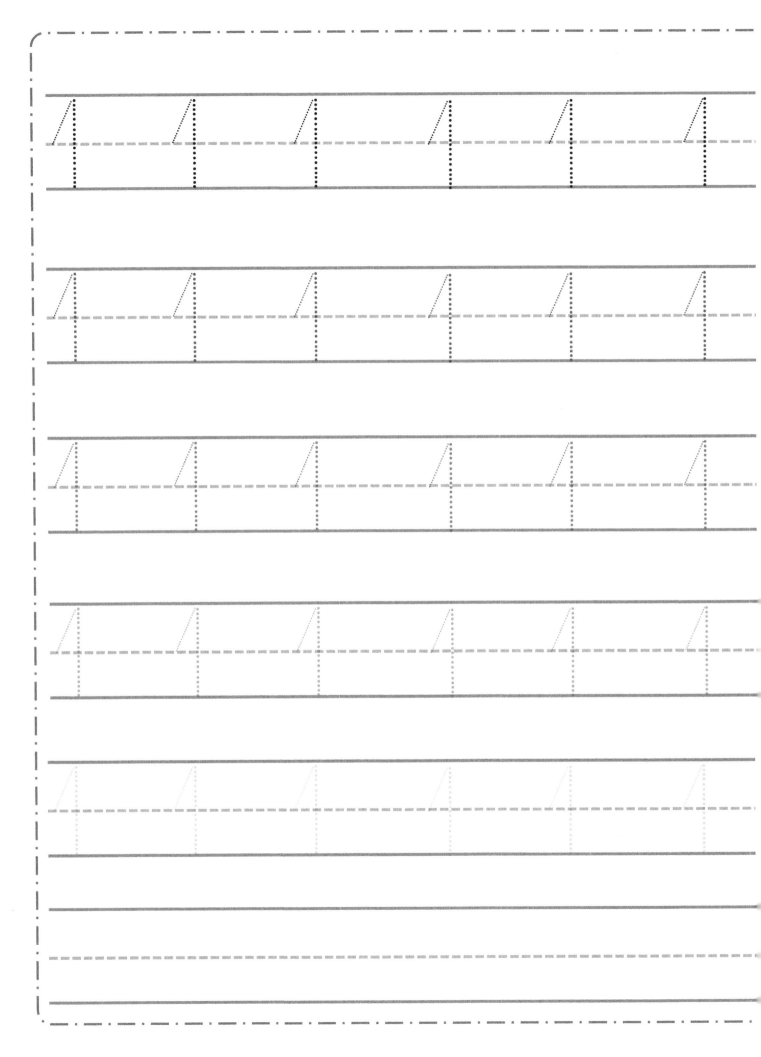

2 Two

2 2 2 2 2 2

2 2 2 2 2 2

2 2 2 2 2 2

2 2 2 2 2 2

2 2 2 2 2 2

2 2 2 2 2 2

3 Three

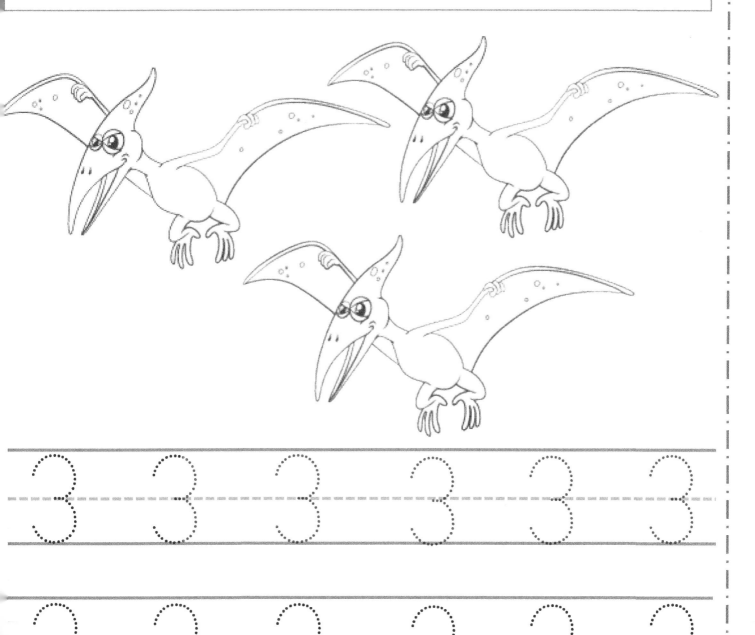

3 3 3 3 3 3

3 3 3 3 3 3

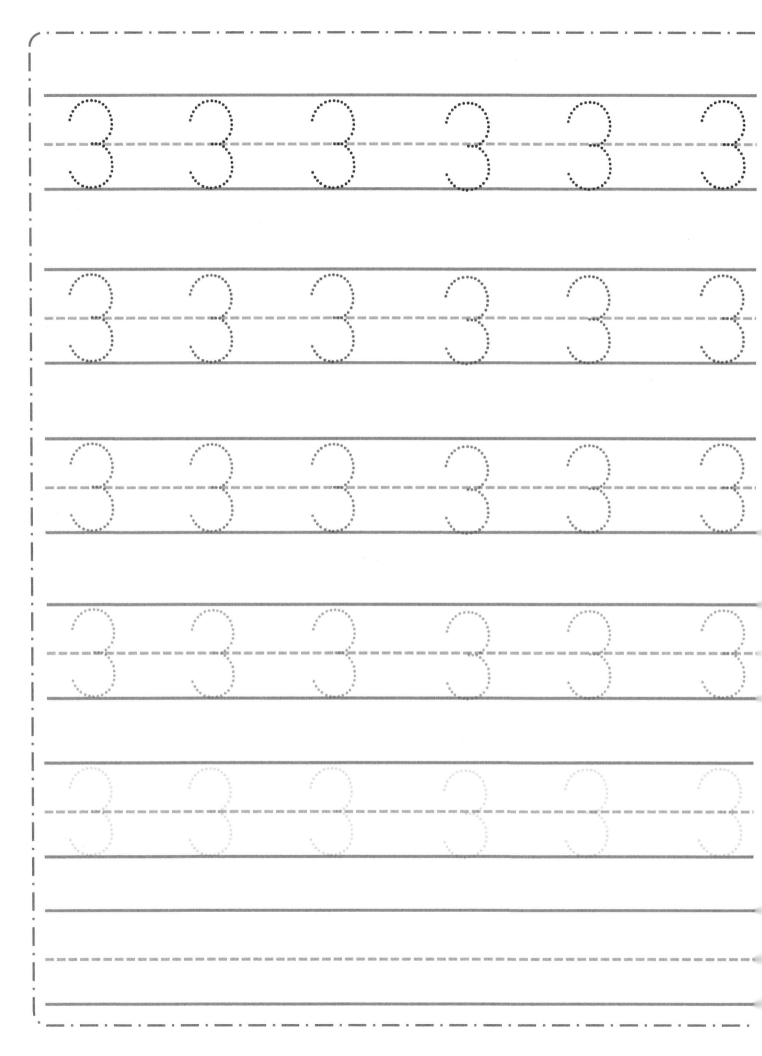

4 Four

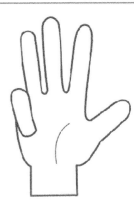

5 Five

5 5 5 5 5 5

5 5 5 5 5 5

5 5 5 5 5 5
5 5 5 5 5 5
5 5 5 5 5 5
5 5 5 5 5 5
5 5 5 5 5 5

6 Six

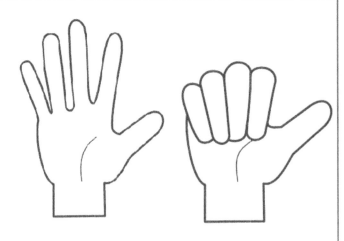

6 6 6 6 6 6

6 6 6 6 6 6

6 6 6 6 6 6

6 6 6 6 6 6

6 6 6 6 6 6

6 6 6 6 6 6

6 6 6 6 6 6

7 Seven

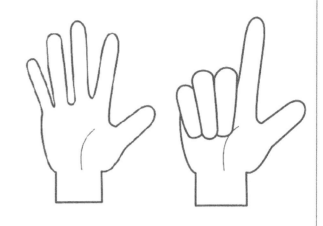

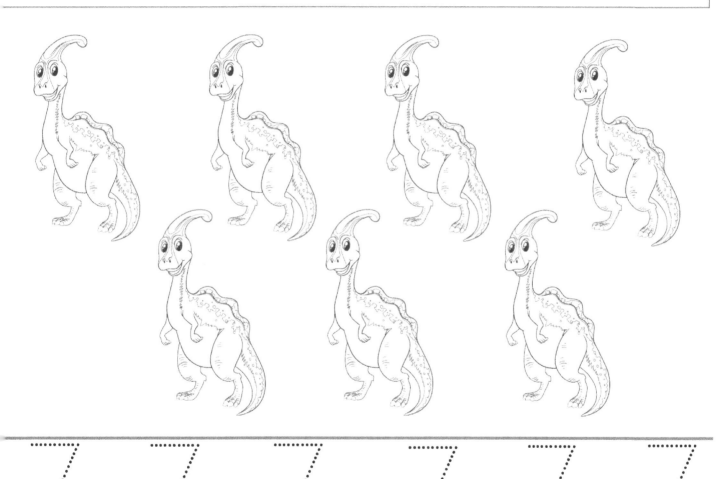

7 7 7 7 7 7 7

7 7 7 7 7 7 7

8 Eight

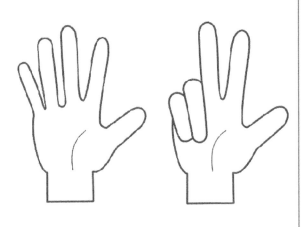

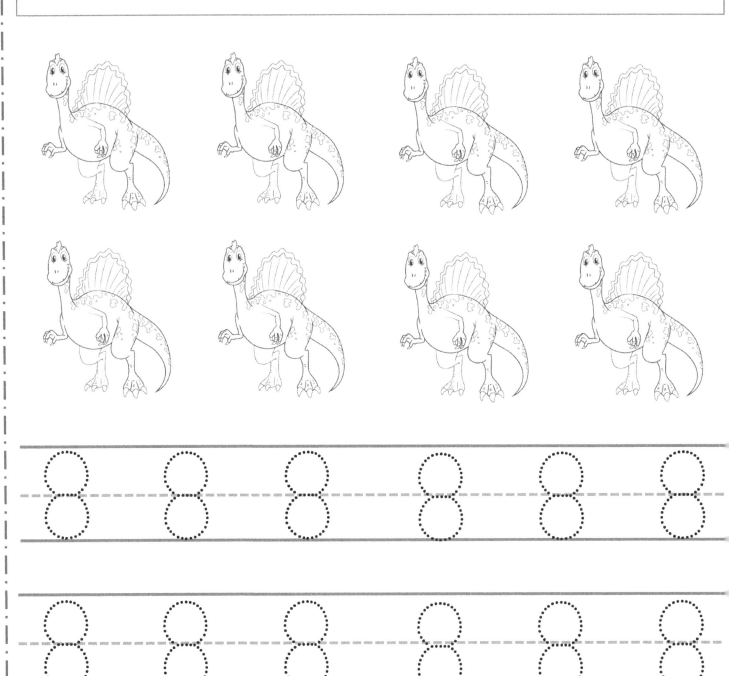

9 Nine

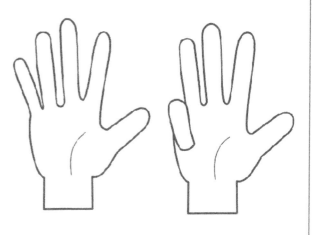

9 9 9 9 9 9

9 9 9 9 9 9

9 9 9 9 9 9

9 9 9 9 9 9

9 9 9 9 9 9

9 9 9 9 9 9

10 Ten

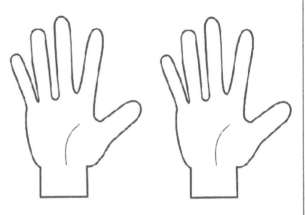

10 10 10 10 10

10 10 10 10 10

10 10 10 10 10

10 10 10 10 10

10 10 10 10 10

10 10 10 10 10

10 10 10 10

ADDITION

Count the dinosaurs

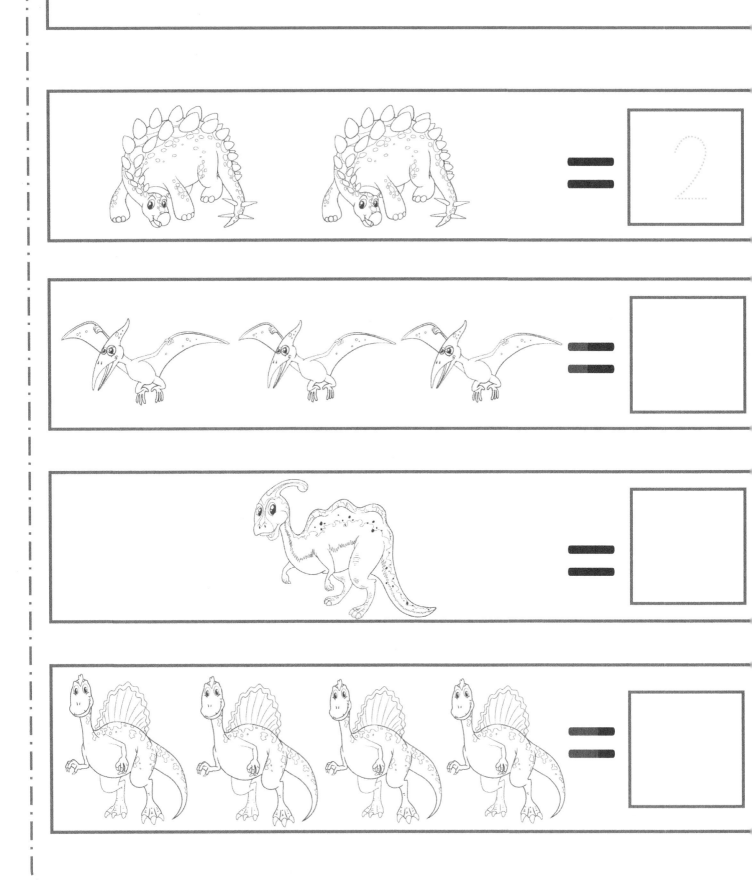

Count the dinosaurs

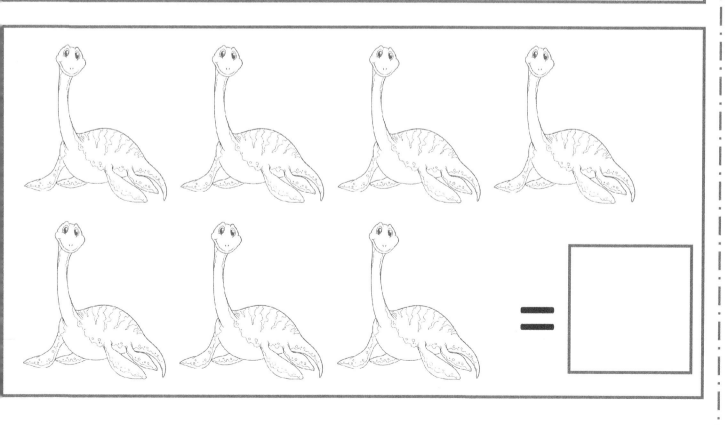

Count the dinosaurs

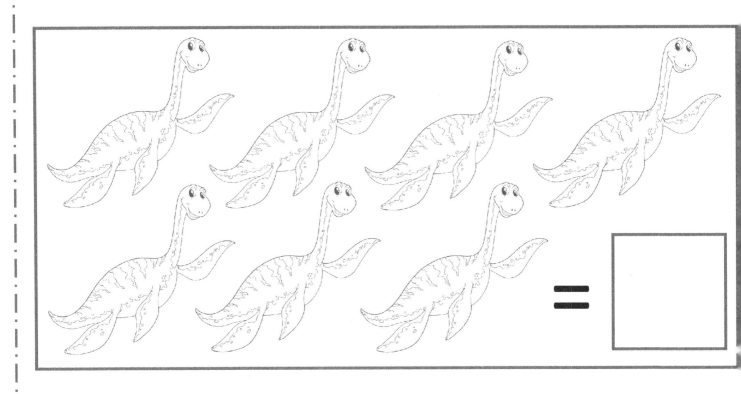

=

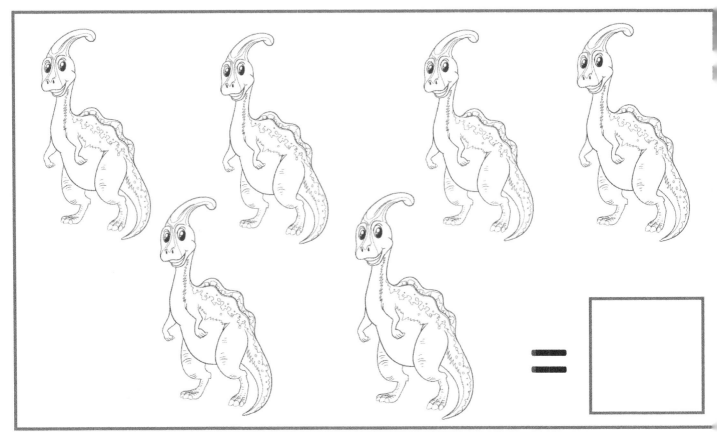

=

Count the dinosaurs

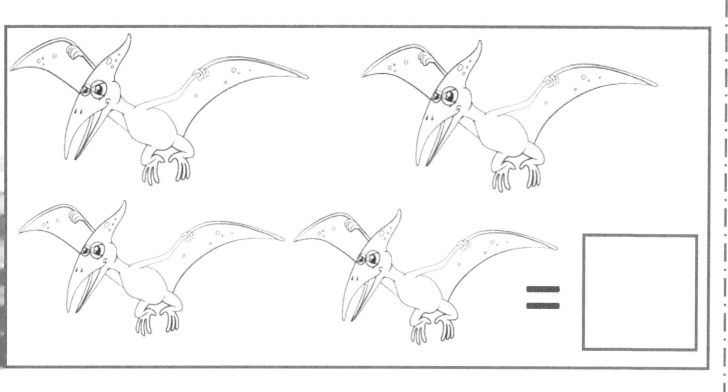

Count the dinosaurs

=

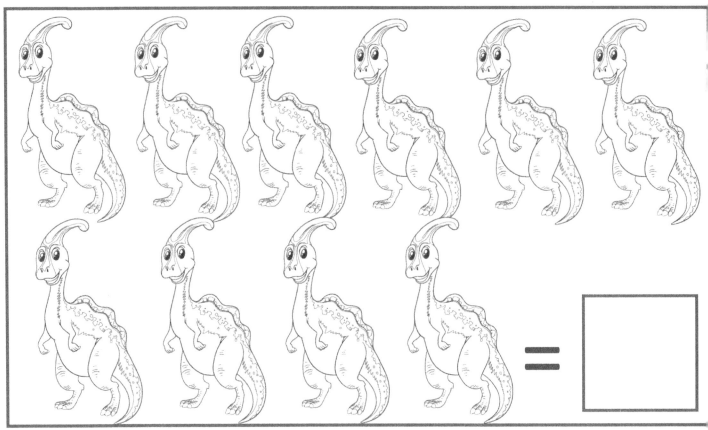

=

Count the dinosaurs.
Cercle the correct number.

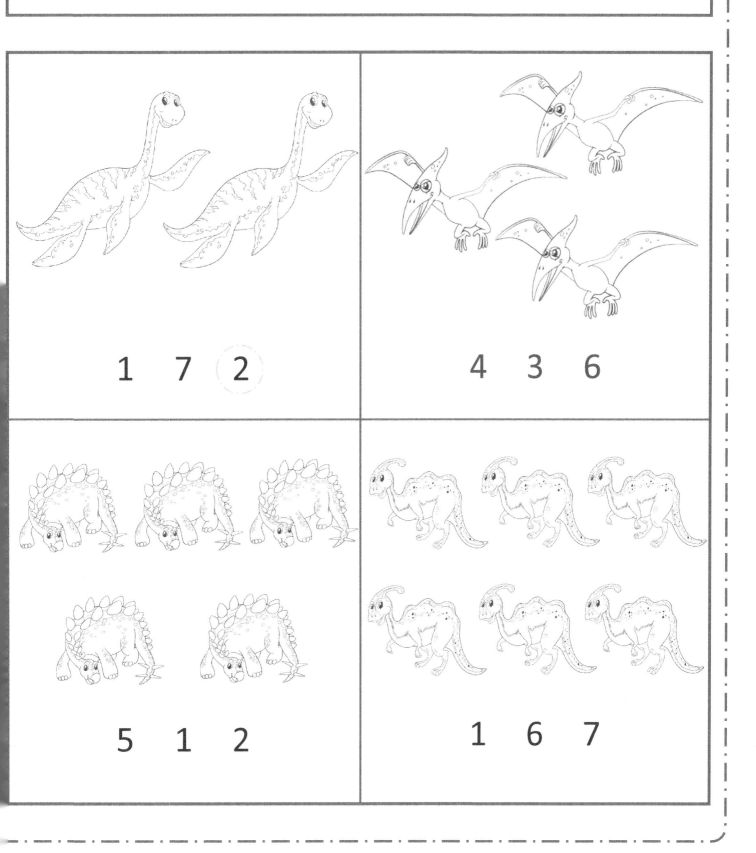

Count the dinosaurs.
Cercle the correct number.

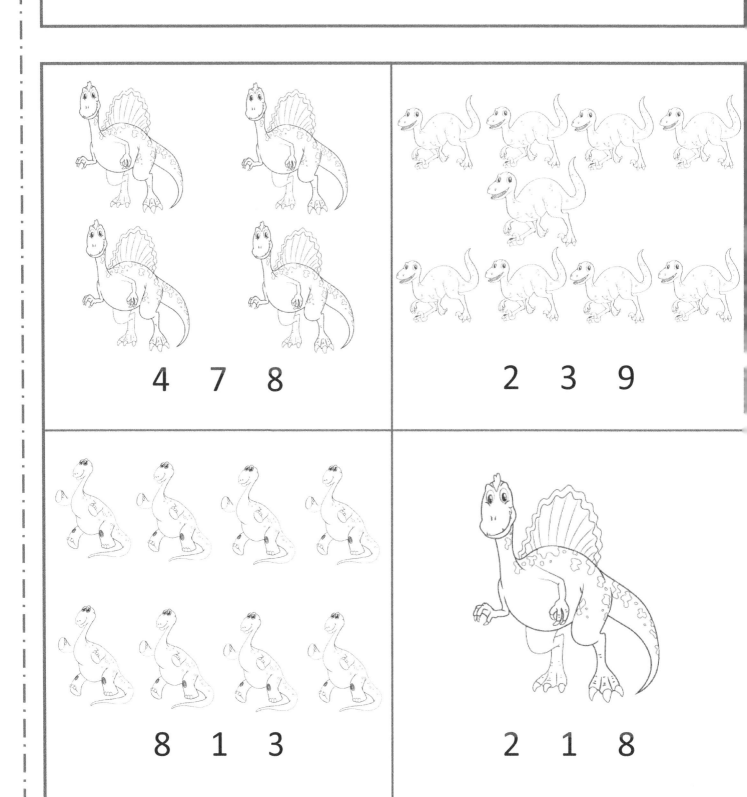

4   7   8

2   3   9

8   1   3

2   1   8

Count the dinosaurs.
Cercle the correct number.

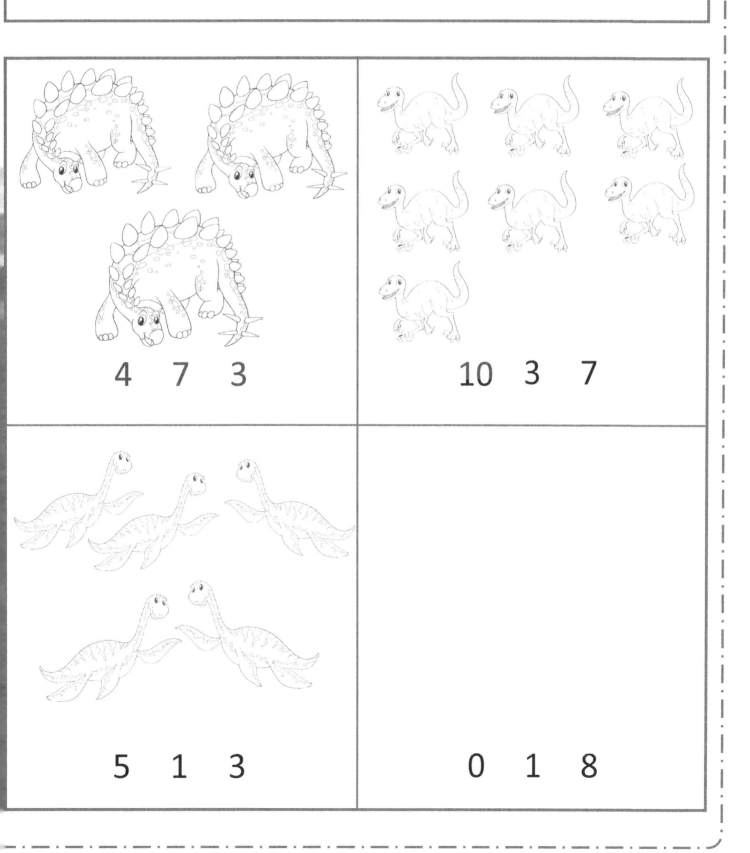

4   7   3

10   3   7

5   1   3

0   1   8

Count the dinosaurs.
Cercle the correct number.

4   6   8

10   1   9

4   1   3

0   1   5

Addition – Level 1

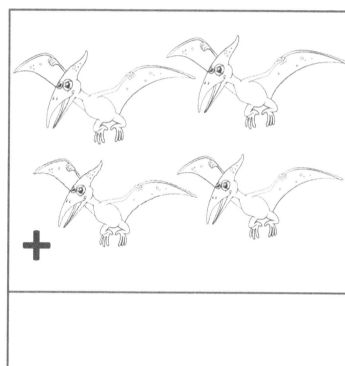

Addition – Level 1

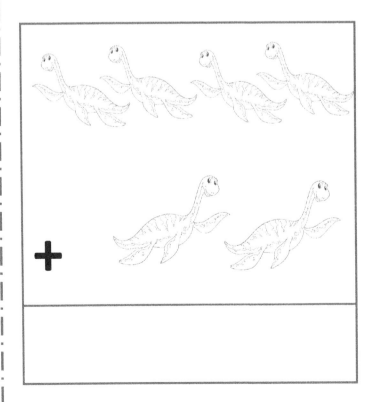

Addition – Level 1

+

+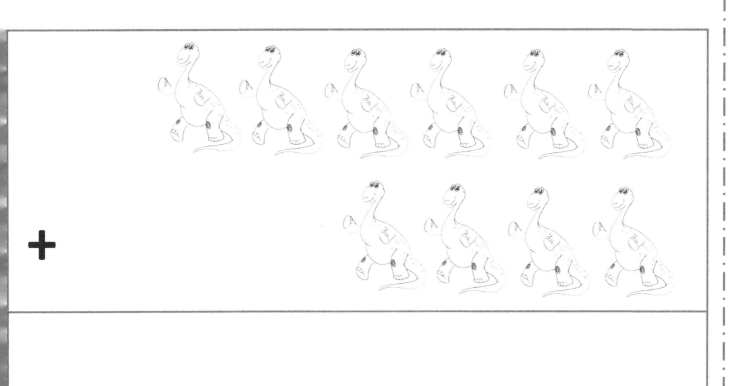

Addition – Level 1

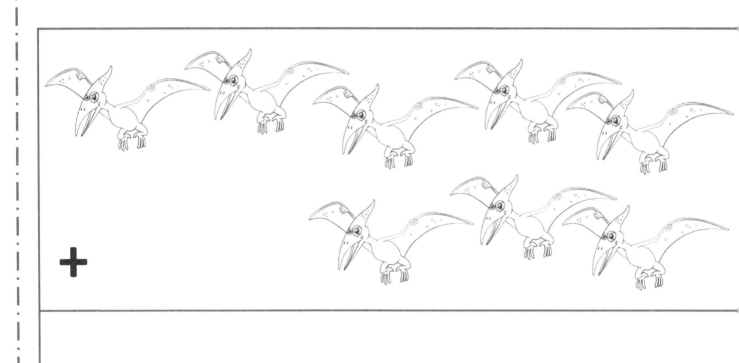

Addition – Level 1

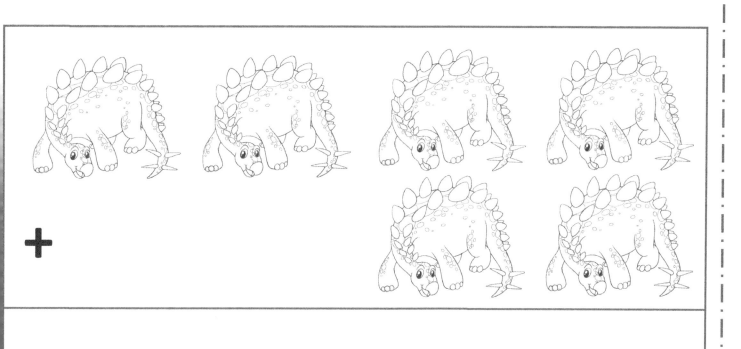

Addition – Level 2

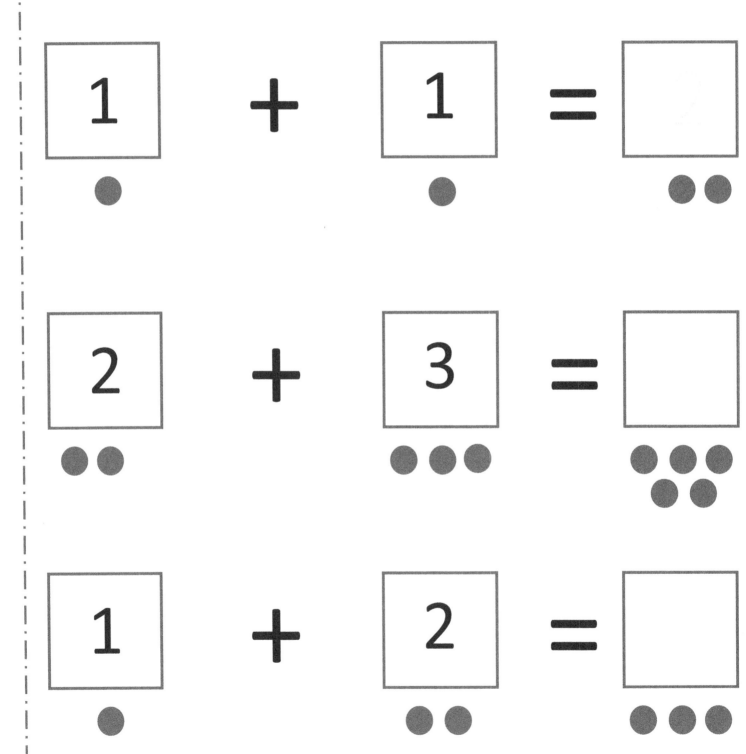

1 + 1 = ☐

2 + 3 = ☐

1 + 2 = ☐

Addition – Level 2

3 + 1 =

2 + 4 =

4 + 1 =

Addition – Level 2

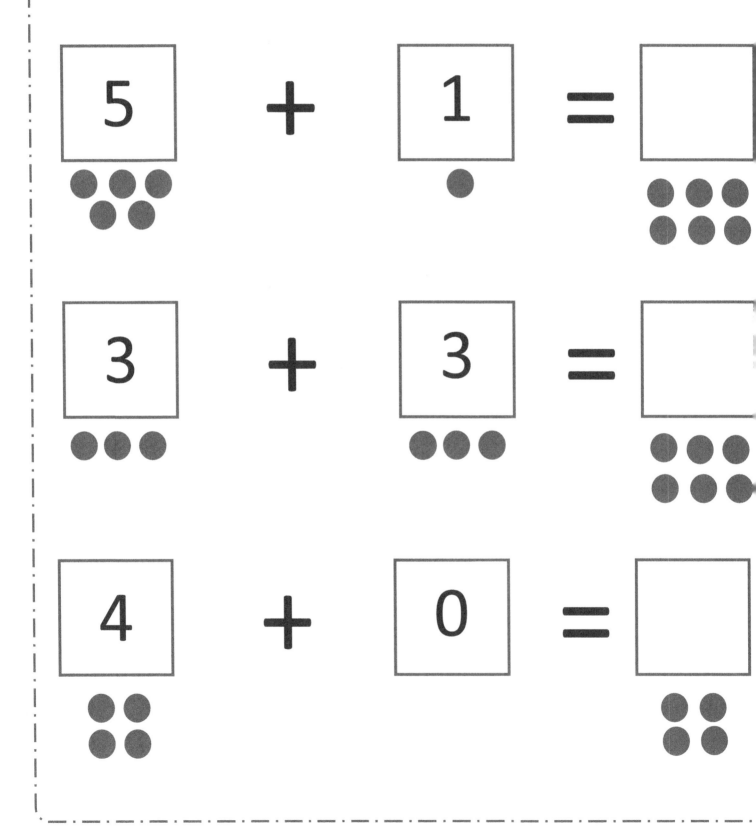

5 + 1 =

3 + 3 =

4 + 0 =

Addition – Level 2

6 + 2 = ☐

1 + 3 = ☐

4 + 3 = ☐

Addition – Level 2

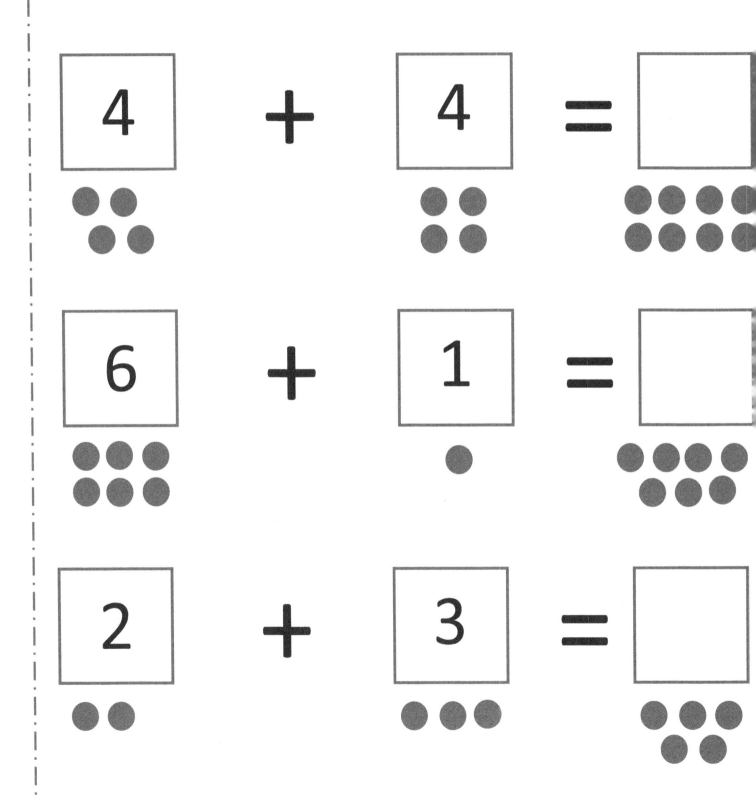

4 + 4 =

6 + 1 =

2 + 3 =

Addition – Level 3

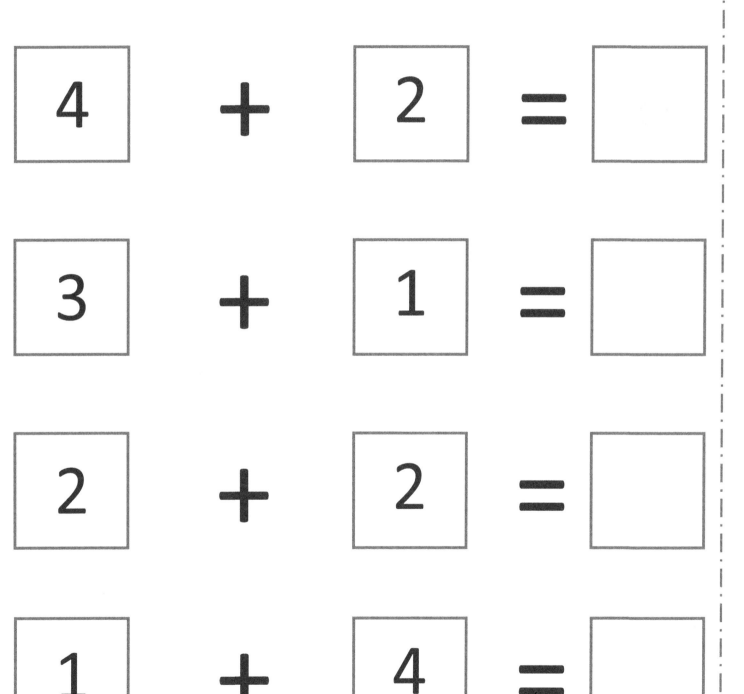

4 + 2 =

3 + 1 =

2 + 2 =

1 + 4 =

Addition – Level 3

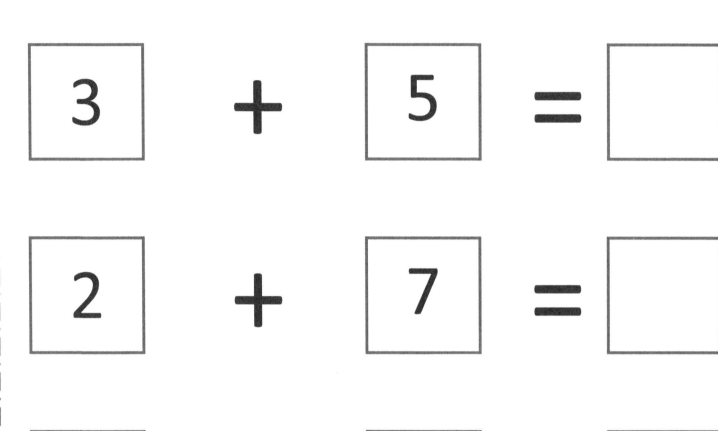

3 + 5 =

2 + 7 =

6 + 3 =

5 + 4 =

Addition – Level 3

2 + 4 =

6 + 1 =

7 + 3 =

4 + 5 =

Addition – Level 3

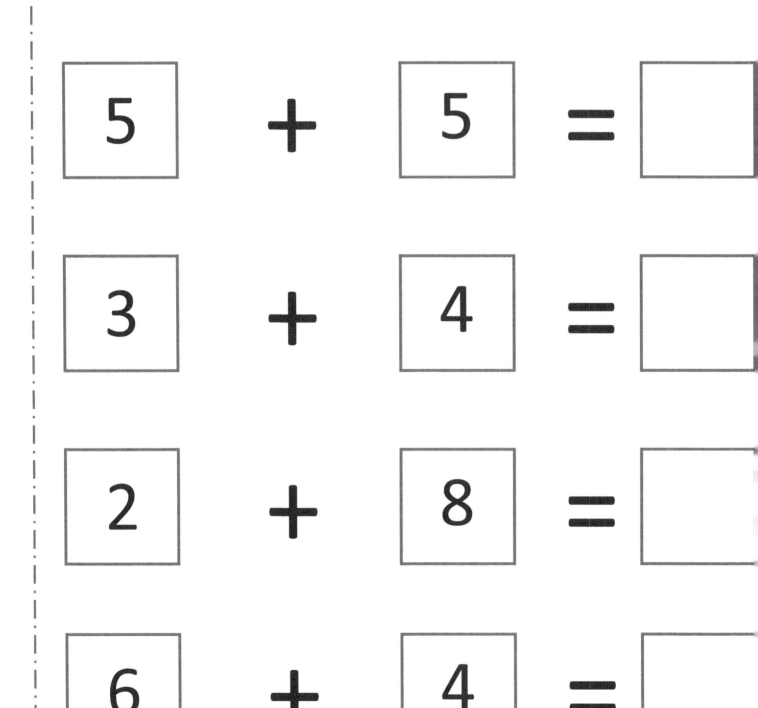

5 + 5 =

3 + 4 =

2 + 8 =

6 + 4 =

Addition – Level 3

9 + 1 = ☐

1 + 7 = ☐

2 + 3 = ☐

5 + 3 = ☐

Addition – Level 3

3 + ☐ = 5

1 + ☐ = 7

☐ + 3 = 8

5 + ☐ = 7

Addition – Level 3

2 + ☐ = 5

☐ + 2 = 4

☐ + 6 = 8

3 + ☐ = 7

Addition – Level 3

☐	+	5	=	5
☐	+	2	=	4
9	+	☐	=	10
7	+	☐	=	9

Addition – Level 3

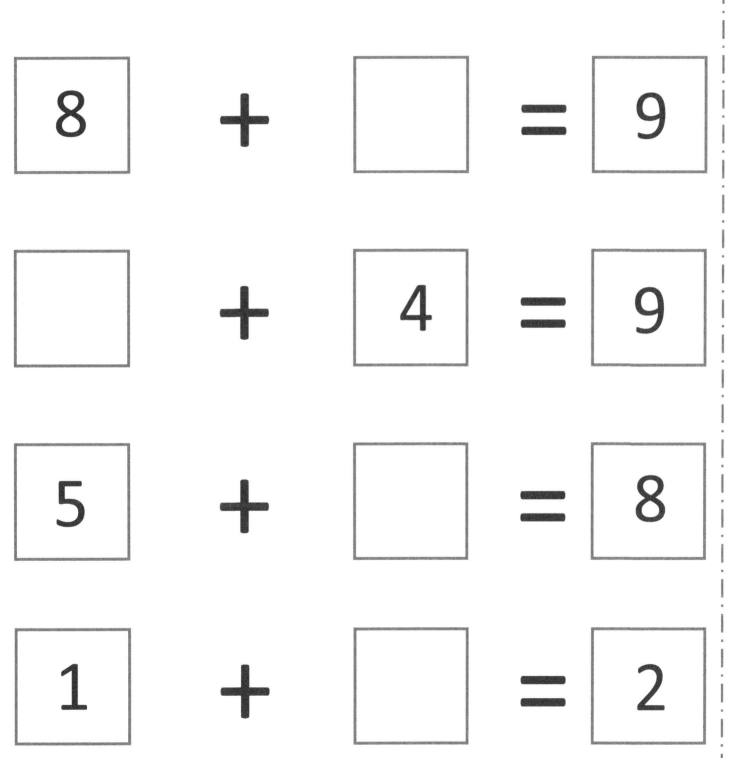

$$8 + \boxed{} = 9$$

$$\boxed{} + 4 = 9$$

$$5 + \boxed{} = 8$$

$$1 + \boxed{} = 2$$

Addition – Level 3

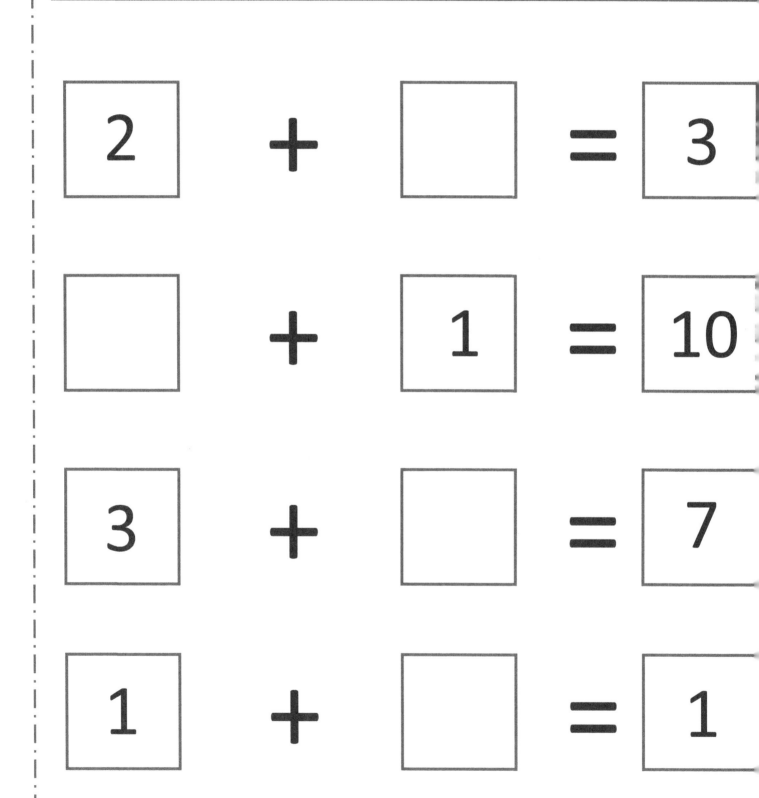

2 + ☐ = 3

☐ + 1 = 10

3 + ☐ = 7

1 + ☐ = 1

Subtraction

Subtraction – Level 1

Subtraction – Level 1

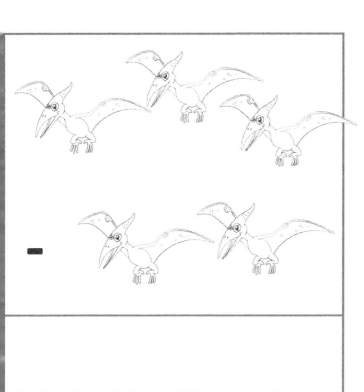

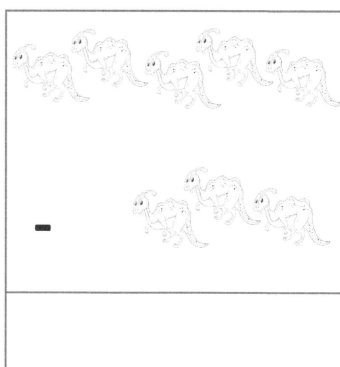

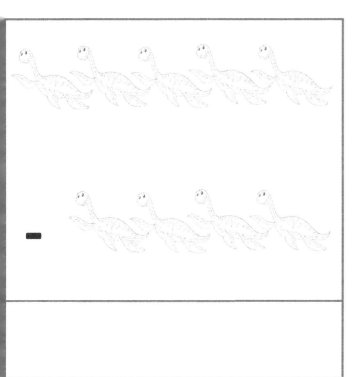

Subtraction – Level 1

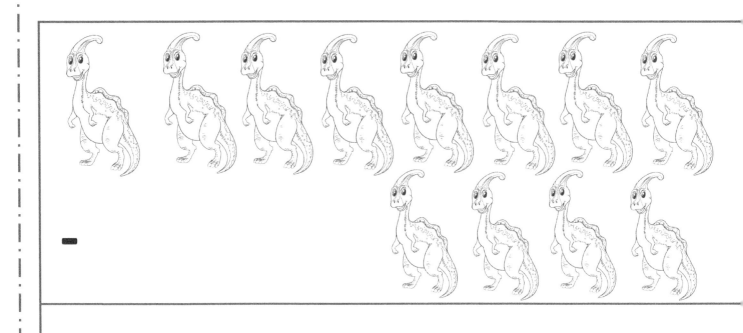

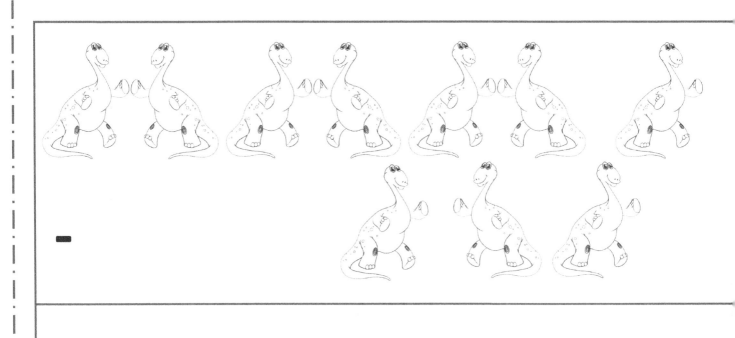

Subtraction – Level 1

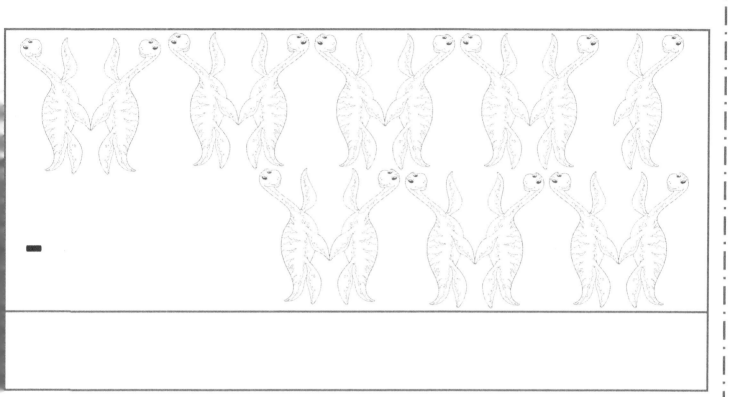

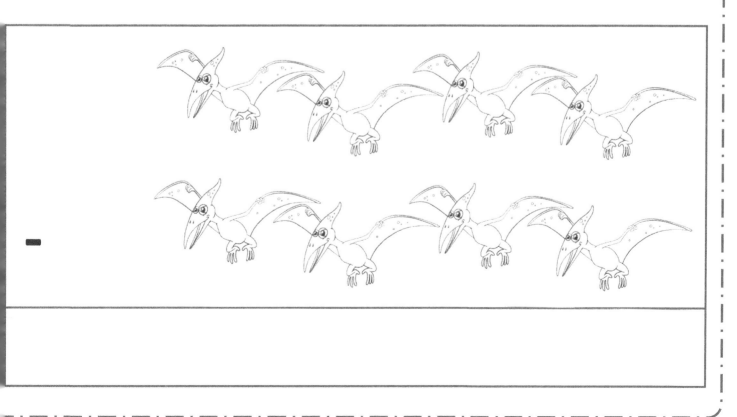

Subtraction – Level 1

−

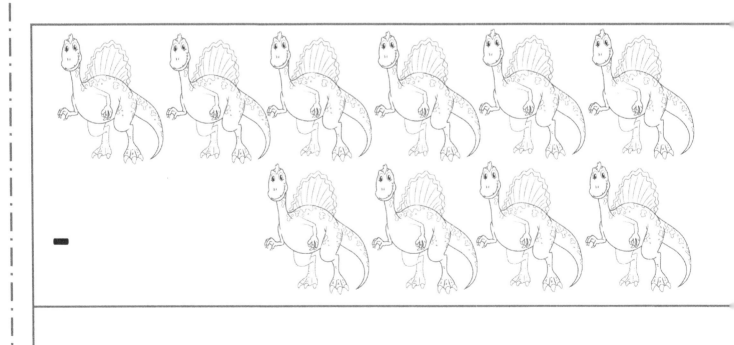

−

Subtraction – Level 2

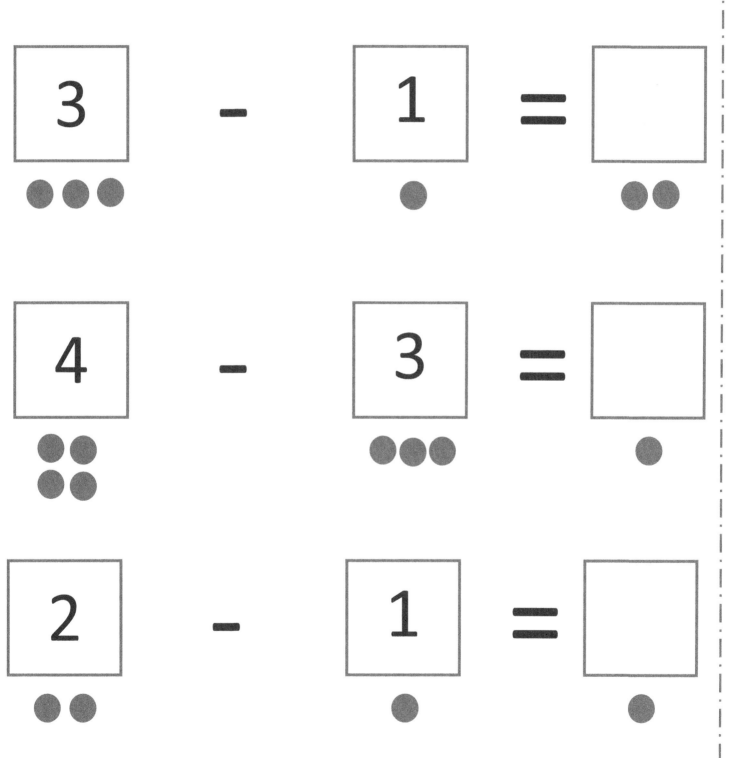

3 − 1 =

4 − 3 =

2 − 1 =

Subtraction – Level 2

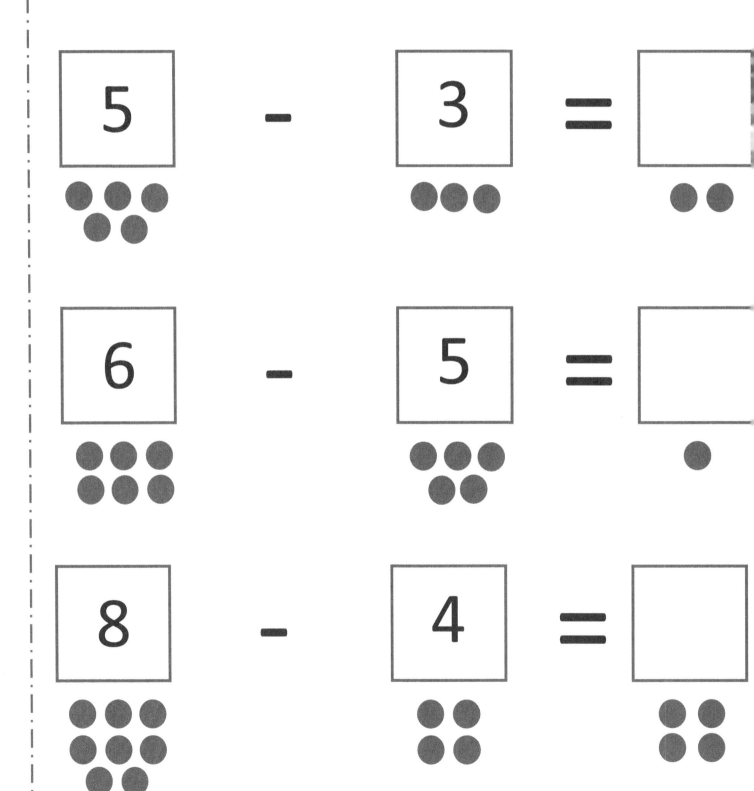

5 - 3 =

6 - 5 =

8 - 4 =

Subtraction – Level 2

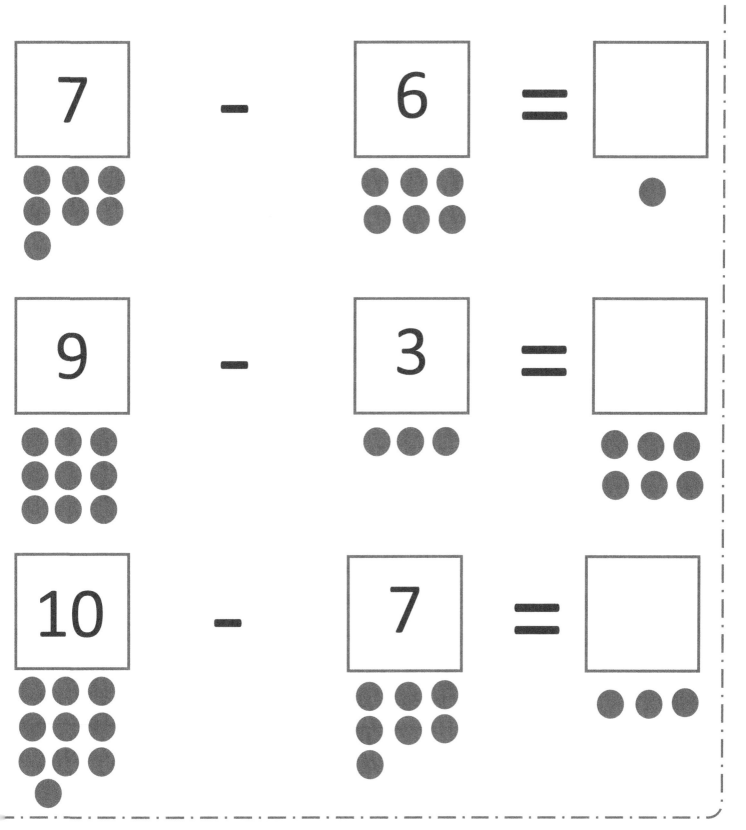

Subtraction – Level 2

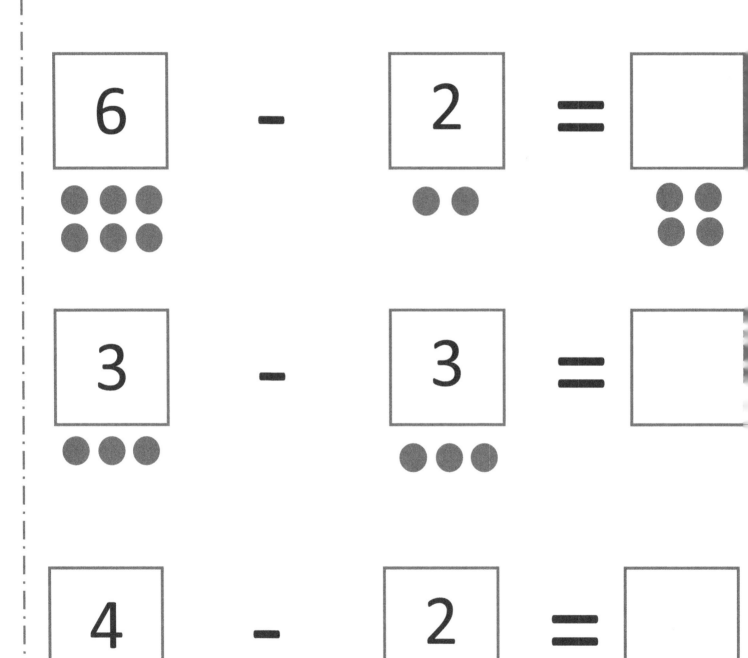

6 - 2 =

3 - 3 =

4 - 2 =

Subtraction – Level 2

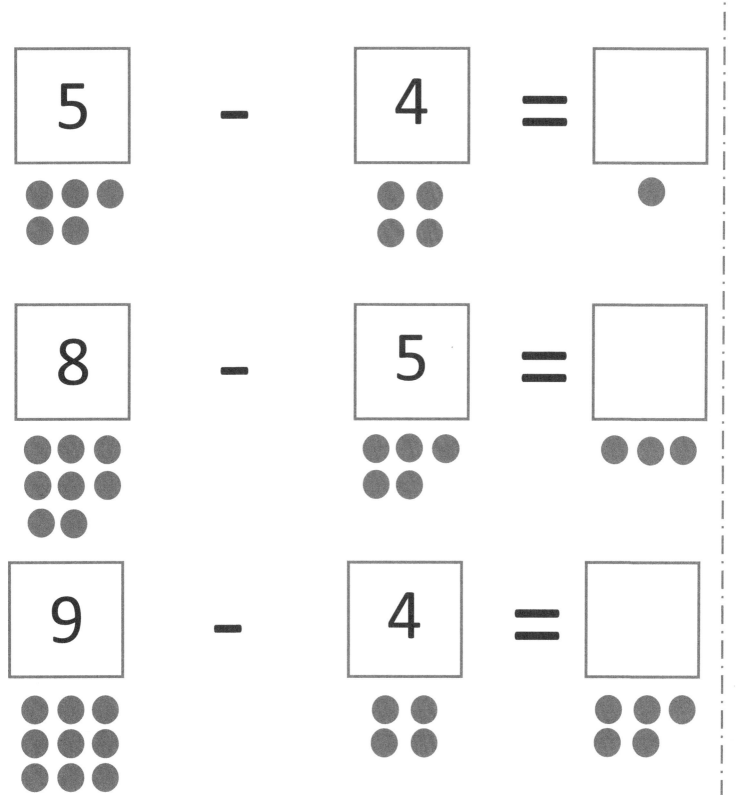

5 − 4 =

8 − 5 =

9 − 4 =

Subtraction – Level 3

10   -   4   =

8   -   3   =

9   -   2   =

7   -   4   =

Subtraction – Level 3

8 − 0 =

6 − 3 =

5 − 2 =

9 − 4 =

Subtraction – Level 3

3 - 2 = ☐

6 - 3 = ☐

10 - 9 = ☐

5 - 3 = ☐

Subtraction – Level 3

8 − 5 = ☐

2 − 2 = ☐

10 − 7 = ☐

2 − 1 = ☐

Subtraction – Level 3

7 - 6 = ☐

6 - 5 = ☐

4 - 3 = ☐

10 - 8 = ☐

Made in the USA
Las Vegas, NV
16 October 2024

96947633R10044